Solar Panel Improvement: 1950-2016

For Solar, By Solar, To Solar

Revised

Begum Fouzia

Y 2025

DEDICATED TO ALL SOLAR ENERGY
RESEARCH PROFESSIONALS

TOC

PREFACE

This book enlisted scientific innovations on solar energy (SE). From solar panel manufacturing information to easy access to variety of accessories for quick installation and maintenance all news are part of this book now. I follow solar energy (SE) development news with great interests for a long time– for every bit of policy changes on SE, for any new discovery news and for all technological breakthroughs, I felt right away to get involved with it. I intended to reach out to general public's door with trust. To motivate them for green energy materials. I collected all solar energy related news from available all sources online, offline or from sideline positions and tracked the trends, rules and policies on SE from my standpoints. I noticed customers' enthusiasm for switching to SE resource and to get introduce with solar tools. Every breakthroughs and technological jump for solar energy conversion have been documented here authentically as well. Thousands of new solar products are in the market now. Si-solar panel is 22.8 % efficient and its comparative lower cost motivated our neighbors to cover their rooftops with these panels to extract his/her own energy.

I advocate for renewable energy. My hands on experience from a US University Solar Energy Research facility helped me to get involve with this current mission. I attempted to synthesized conjugated organic compound to use as light capturing material for solar panel in the lab. Project's requirement eventually exposed tons of advance research articles on SE to me – exposing the huge potentiality of this field and the opportunities in this domain. While in the lab, in carrying out day to day analysis I sought quick alternatives for synthesis, however, final product characterization was tremendously problematic and slow paced. Lack of people with expertise to operate the sophisticated instruments obstructed my entire projects'

outcome.

Y 2016 is past now. In third world countries years just fly

byes for people. But in advance world, we see often new tall

skyscrapers facades that replaced with photovoltaic windows

on its shiny corners. Or, houses with rooftop panel and

installed storage cells to better manage resident's energy

pursuits. Even so, cumulative conversion from all these solar

initiatives only accounts for ~ **1%** of the world's electricity,

though overall, solar growth's standing now is **12%**

globally. Environmentalists labeled this 1% as **zero-carbon**

electricity source and opined that **a race** is needed globally

to attain a higher percent response. Most convincing

part in solar drive today is, once panels are installed and

some sort of contract is signed off, the system then is paid

off for the next 20-25 years. During this long period panels'

generate power continuously for free. All news on

innovations on solar research and development ideas and live demonstrations, helpful links with videos registered in the book to make the search handy for SE. The book's core objective is to provide on site help during installation schedule for everyday people. My pledges are:

> **i) Track down every incremental changes of solar panel, and for other devices in this industry.**
> **ii) To demonstrate easy to understand solar technology installation and maintenance steps.**
> **iii) To produce a handy book written in easy language and terms that be available to all nearby book stores.**

When enough people actively get engaged in the pursuits of any cause the outcome turn more meaningful. The book is for people of all cast, creed and culture and our campaign slogan is:

For solar, by solar, to solar.

CH - 1

INTRODUCTION: SOLAR CELL
History, Components & Design

Scientists or non-scientists, people worldwide are racing to find renewable energy as alternatives to fossil fuels such as coal, gas and petroleum. Fossil fuel is also at least 3 times less expensive, but a serious carbon polluter to the environment and its reserve is shrinking gradually. We, the conscientious people's group see huge potential in sunlight which is the most abundant and cleanest form of power source to be used as fuel. Searching for the finest recipe for low-cost, light absorbing materials to harvest sun light to produce energy is now the primary consideration not only for people in developed countries only, rather it is now a global dream. Silicon Solar panel manufacturing cost is very high.

Energy is required to produce a silicon **ingot** (an oblong shaped block of silicon), unless we cut down this consumed energy cost, Si-PV will not be considered a reliable form of energy source.

Discovery of Solar Energy & Invention of Solar Cell

In **1839, Edmond Becquerel first** discovered the principle behind solar energy known as Photovoltaic Effect.

While working in his father's lab, Becquerel noted, some materials can generate **voltage (V) and electric current** (Amp) when exposed to sunlight. This **physical and chemical** phenomenon is called the **photovoltaic effect**, known also as the Becquerel effect. To demonstrate this photovoltaic effect, Edmond immersed a platinum and a gold plate together but separately in an **acid, neutral, and alkaline** solution and then exposed the set-up to solar radiation. His demonstration created the first instance of an 'electrochemical cell but was not very efficient.

Years later in **1873**, Willoughby Smith (an English scientist) observed that **Selenium** (atomic number, Z: 34) is a very effective material to create charge upon light hitting.

Then in **1881**, Charles Fritts (an American scientist) **created the first solar cell** (first solid-state photovoltaic cell) using **Selenium coated with thin layer of gold**. That solar cell was only around **1% efficient** at that time, but it was a huge discovery. Using Selenium Fritts designed the very first solar array that was installed on a New York City rooftop using selenium cells.

In 1905, Einstein published his first paper on **"Photoelectric effect"** elaborating the detail mathematical aspects of solar radiation, how it works and its threshold values.

Modern Solar Panels in use Today…

Russell S. Ohl was an American **semiconductor** researcher and an engineer by profession. In 1939, he discovered the **P-**

N junction (a positive side and a negative side inside a single crystal semiconductor). This discovery led him to create the **first silicon solar cell in 1941**. The design of this silicon solar cell (**Russell's version**) still used in today's modern solar photovoltaic (PV) panels!

Solar cells are the building blocks of **solar module** and **solar panel**.

Commercial solar photovoltaic industry reached 63rd birthday this year (2016). Most modern solar cell idea was kicked off in 1954 by Bell Lab [Note - 1] scientists. The solar industry was nonproductive at that time. In the 1980s, a new prototyped solar cell was developed for uses. That device also couldn't make people happier instead doubled down skepticism on SE. Issues noted at that point were solar cell's inefficiency, its greater installation cost, unreliability, and the requirement of strong government subsidies to holdup. So, research continued in this field, at the end SE converted to a reliable form of renewable energy source. At that point, critics' all concerns on solar energy proved just plain wrong. The fact that sun as the primary energy source is ubiquitous, and we do not spend a single-dime to buy this fuel anywhere, all investments encompasses only the set up costs for a device to extract, convert, to store or to operate appliances at home. Based on expert's calculation just 1 hour sun shine in a sunny day is enough to power the entire world for one year. Production cost of 1hr. solar energy using Si-PV panel is area dependent, larger the area to cover, higher would be the cost.

[calculation is showed in **CH - 3 / Solar Energy Terms.**]

Photovoltaic effect is the generation of voltage (V) and electric current (Amp) in a material upon exposure to light. It is a combination of physical and chemical phenomena, fundamental steps of which are: Light energy absorption causes excitation of an electron or other charge carriers to a high energy state. Excited charge carriers still retain within the material using diffusion kind of process and the voltage (electric potential) is produced due to the separation of charges. The light energy that absorbed must possesses sufficient energy to overcome the band-gap barrier for excitation.

Photo electric effect is upon exposure to light an electron is ejected out of a material (usually into a vacuum).

A **solar cell is a photovoltaic cell**, is an electronic device that converts energy of light directly into electricity. Its electrical characteristics such as current (I,amp.), voltage (V, volt) or resistance (R) vary upon exposed light. A single P-N junction solar cell can produce a maximum **open-circuit volage of ~ 0.5V.**

Global population is increasing and so is the demand for

total energy uses. Public in general is keen to know more about cheaper, non-toxic, abundant and efficient solar power so that they can reliably switch to green energy resources. And for scientists' issue now is not "how to capture sun light," rather "how to convert sun light efficiently?" And for PV industry, it is not how far the industry has progressed, rather how many more miles the industry still has to cover before delivering enough clean energy to mitigate essential energy demands for consumers. Manufacturers' are committed to bring innovative products to the market. By pouring additional resources to R & D sector, companies like **Firstsolar, Sunpower and SolarCity** are bringing innovations to the market. Traditional silicon panels are blue or black. Universal solar energy conversion scheme (aka. SPISMA) is presented here:

(Credit: https://www.fbsolarllc.com)

Figure 1.1: Solar Energy Conversion Scheme: SPISMA

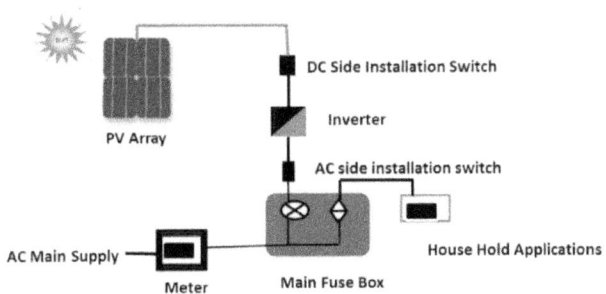

Fuse Box
↑
Sun→ PV module → Inverter → Switch →Appliance
↓
Meter

Description: Photons from solar radiation knock electrons to loose from the photovoltaic material. Electrical conductors in the circuit capture the electrons in the form of electric current that is 'electricity.' In the circuit- liberated electrons pass through **inverter/switch/fuse** finally to the meterbox and appliances.

SE Insights: Photons (hv, package of energy) reaching the semiconductor surface will be either, i) reflected from the surface (R_F), ii) or, will be absorbed in the material (A_A), iii) or will be transmitted through (T_T). For any solar cell R_F and

T_T paths are considered loss mechanism, only A_A path is effective and create energy. No photon absorption, no electricity production. Absorbed photon has the ability to kick out an electron from the valence band (V_B) state to conduction band (C_B) of the semiconductor unit. Energy difference between the two states V_B - C_B is called **band gap energy**, $E_{BG} = (V_B - C_B)$. One electron ejects in this process. If E_{hv} is the photon energy: it can create three possible scenarios:

i) $E_{hv} < E_{BG}$; no light/electricity generate,

ii) $E_{hv} = E_{BG}$; creates a hole (+) and an electron (e⁻) pair

iii) $E_{hv} > E_{BG}$; allows absorbed energy to create heat

Electrons from photovoltaic material travel through the circuit is the direct current (DC) — essentially a torrents of electrons all flows in the same direction— be required to pass through inverter. Inverter splits the electron flow rapidly flickering between two ends of the circuit. Two streams of direct current (DC) meet head on in a transistor (switch) through wiring and turns to crude alternating current (AC) as AC source. Other devices connected in the circuit are simply to run the device efficiently, to amplify the signal and to store and monitor charge.

A commercial version of solar cell has been constructed below. Two thin Silicon PV wafer doped chemically. Based on the chemical agents silicon wafers turned either to p-type or n-type semiconducting layers upon connection in circuit yield solar energy (SE). Net cell design remain identical to SPISMA model everywhere. Here is the finished product of a modern commercial solar cell:

Figure 1.2: **Turning Rays into Energy-Harnessing the Sun**

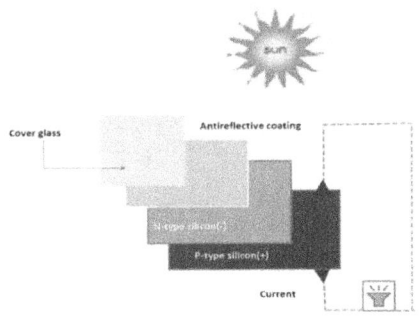

Construction - **Two oppositely charged semiconductor layers placed atop with n-type above p-type, loosens electrons (upon light energy striking the cell). Loosen knocked electrons can be captured through the electrical conductors attached to the positive (+ve) & negative (-ve) sides of the electrical circuit to generate direct (DC) current. An inverter (not shown in the figure) turns DC current into usable alternate current (AC) of electricity to use and for to modify.**

Applying research and innovative ideas solar industry is surviving and panel price today has dropped (almost 80%) standing below $1.0 per Watt. Efficiency remained to be the primary concern for this industry. At present, Si (PV) panel efficiency is 22.4%. A great deal of light energy is lost as heat in the process.

INSATLATION GUI DELINES

Pre-Installation Steps: #Sign up for renewable energy (solar) installation program, #Get approval from GRID Alternatives, #Calculate how much power is needed for a home.

Requirements:

- Desert area (no shadow), rooftops or, any open space
- Solar photovoltaics made of: Si-PV, or from Thin-Film Technology, or, OPV (Organic PV)
- Metal frames or racks to hold panels positioning sun facing tilted (37^0) to the surface
- Master cable, clip, **Inverter** – all accessories in line

On Site Work—begins with climbing ladder and grabbing a drill:

> •Carry the panels to a staging area in the safe vicinity,
> •First, from roof top crew members drop pair of ropes to clip the panels to hoist up,
> • On roof top panels get connected with the racks (installed earlier), adjusted turning the cover glass correctly,
> • Wires be clipped into master cable,
> • Turn on the switch.

<u>Activity</u>: Panels mounted on the roof in a tilted position (hotter than the ground). Few inches of space must remain free between the panels and roof so that air flow can easily pass through and cool the panels down naturally. Panels absorb sunlight creating oppositely charged carriers-electrons (e⁻) and holes (+) on silicon surface. The charges must move away quickly (in the order of a pico second, 10^{-12} s) in order to produce electricity. Failure to move or to separate the charges fast enough ($>10^{12}$ s) results in recombination of charges and effectively no change occurs. The overall **efficiency** of solar panels compares how much

recombination occurs and how many charge separates. Efficiency itself is judged by three parameters, <u>voltage</u> of open circuit (V_{OC}), <u>current</u> of short-circuit (I_{SC}) and <u>Field Factor</u> (FF). High temperature disfavor solar PV panels work. Heat causes electrical resistance to the flow of electrons. Resistance in electrical wires consumes power. Power consumes increases as current pass through the wire surges. High voltage at two ends of a wire ultimately reduce the current passing through it turning electrical system inefficient. At or above $75°$ C, the electrical resistance makes

the voltage (V) fall thereby producing less kilowatts hour (kWh) of energy. The best solution: **N**atural **C**ooling

EFFICIENCY- DETAILS

All solar cell efficiencies are measured at STC unless otherwise stated. STC stands for standard parameters of $T=25°$ C, an irradiance $1000W/m^2$ with an air mass (AM)

=1.5 %. In standard condition Silicon solar panels demonstrate only ~30% efficiency. This value is an absolute theoretical limit on traditional solar cell efficiency for energy conversion. The value was determined with only **one single-layer of silicon cell** which demonstrated the upper limit of efficiency only 32%. By using multi-layer silicon cells effort to increase efficiency (η) has not been succeeded yet. Overall, by measuring ratio of electrical output of a solar cell to the incident energy in the form of sunlight and its relationship with temperature (T) which is inverse [means, if **T is↑, η is↓**], efficiency η is calculated out. In solar energy conversion process, incoming photons are considered as one single entity which is the white light consists of all different wavelengths (λ) of light (way shorter $<\lambda$ to way bigger$> \lambda$). In this mechanism a large portion of light-energy get wasted reaching upon the panel surface because low-energy light particles do not get absorbed from the incident rays due to lack of adequate energy to bridge the band gap of the crystalline-Si solar cells. On the other hand, high-energy photons upon incidence get absorbed, but in just a few picoseconds (10^{-12} s) interval of time much of their energy get transformed into heat damaging the cell. Study revealed, efficiency goes down by about **0.5 %** for every increase of T by **1.8^0 F (1^0C)**.

The wasted heat heats up solar cells to 130 ^{0}F (55 ^{0}C), or even more. This makes Panels' exposure to sunlight ineffectual. Due to this ongoing dual phenomena only a small portion of the solar energy get convert to electricity by PV cells limiting maximum efficiency to just 30 percent. Panel efficiency and other conversion equations in solar energy displayed below:

$$\text{Efficiency, } \eta = (\mathbf{P_{OUT}}/\mathbf{P_{IN}}) \times 100\%$$

[where, $\mathbf{P_{OUT}}$ is panel output Power and $\mathbf{P_{IN}}$ is incoming irradiance]

X% efficiency*1000W/m2*Area (m²) = Power (W) =

panel output

$$PanelPower(W) = X\% \ (efficiency)1000W/m^2$$

$$Power(W) = X(panel \ output \ volage)\frac{W}{m2}$$

$$* \ Total \ Area(m2)$$

$$Energy(kwh)$$

$$= Area * effici. * Total \ Irrad. * hr.$$

$$* \ System \ loss$$

E = a*r*h*PR	E = Energy generated a → area, r → solar radiation (η*1000W/m²) h → dùration, PR→ is for system loss]

[Energy calculation based on this equation demonstrated in CH - 3 of Solar Terms.]

Society today is better skilled, informed and trained for solar installation – the endeavor is continuous. Today, we do have better diagnostics tools, better ways to reach customers directly and better solar accessories for easy installation — research on all branches helping constantly to decrease the solar energy cost. From scientists, researchers, engineers and policy developers' to vendors, suppliers, technicians to simple roof-top home owners all are the players in this great transition. Their combined effort is adding improvement in solar industry inch by inch. Now, with PV storage and backup power option for the worldwide residential solar market, along with efficient inverters use, solar power system is getting maximized for an individual PV module owner while lowering cost of energy in the first place by solar PV system. Use of aluminum racking system lowered the price further. Overall, renewable energy technology is evolving slowly but in steady pace. It takes even decades to full blown a system when things go well.

₪₪

FootNote: Transistors — are electrical switches that control how electricity flows around a circuit — unable to withstand the high radiation environs around.

CH- 2

PANEL MANUFACTURE
SOLAR PHOTOVOLTAICS

Solar panels are made up of silicon photovoltaic (PV) cells.

When sunlight hits solar panels, the solar PV cells absorb the

sunlight's rays and electricity is produced via the

Photovoltaic Effect. The electricity produced by the panels is

called Direct Current (DC) electricity, which is not suitable

to be used in our home for appliances. Instead, the DC

electricity is directed to a central inverter (or micro inverter,

depending on the system setup). Silicon is a semiconducting

material. At first "p-type silicon" was used as standard

material for cell making, switched now to "n-type silicon"-

that lasts longer than the p-types.

2.1: <u>Silicon Panel (Si - PV) Manufacture</u>

Elemental silicon (Si_{14}) exists in amorphous (a-Si) and crystalline (c-Si) forms, and in between the two extreme ends exists partial crystallized silicon. The partially-crystalline silicon is often called polycrystalline or Poly-Si (or multicrstalline, mc-Si) in short. Polycrystalline form of silicon is also of high purity and used as raw material for solar photovoltaic and electronic industry.

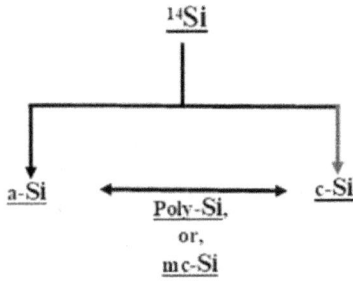

Silicon in Nature

Crystalline (c-Si) silicon dominates the **wafer** based solar cells. At first, thin-film silicon modules were produced from amorphous silicon (a-Si). This technology could only reach 8% to 12% efficiency and even by Y-2020 its record lab efficiency stood at 14%. Because of its reduced silicon intake criteria made this product an attractive option to some manufacturers during the last decade of 2010. Meanwhile, in the market polysilicon price dropped gradually down from ~ $60/kg in 2010 to less than $20/kg in 2013, this driven out amorphous silicon (a-Si) quickly contesting against crystalline silicon (c-Si) cells. Manufactured silicon solar panels (Si – PV) categories follows as:

- **thin-film (amorphous-Si)**
- **monocrystalline (c-Si) (black colored)**
- **multicrystalline (mc-Si), or Poly-Si (blue colored)**

Each type of panel has its own pros and cons. Crystalline solar panels are generally **more efficient** (require less area for the same watts). Thin-film panels **lose less power @ high temperature**. The main difference between the **Poly vs. Mono** silicon is in the crystal purity of the solar cells. Monocrystalline solar panels have solar cells made from **a single crystal of Silicon** while Polycrystalline solar panels have solar cells made from **several fragments of silicon melted** together. Polysilicon solar cells consists of small crystals known as crystallites (creates a flake like effect).

Between Polysilicon & Multi-silicon (the two varieties are often used as synonyms): usually **crystals size is larger than 1milimeter (1mm)** in multi-crystalline type and this curved the most common type of solar cell for most produced polysilicon in global PV market. (mc-Si) solar cells are of **blue** colored.

Most importantly, the higher crystal quality in c-Si cell advances charge extraction and power conversion qualities, but produces wafers that are 20% to 30% more expensive. The active regions of PV solar cells are composed of **>100μm thick wafer** that give and receive electrons to generate electrical current when exposed to light. Relatively thick wafers absorb light strongly weighing about 50 Ib.

About Wafer (How it Produce): A wafer is a-thin slice of semiconductor substrate such as c-Si used in electronics for fabrication of IC circuits and for conventional wafer based photovoltaic solar cells. A silicon wafer is made by spinning molten silicon in a crucible. A small seed crystal is inserted and slowly withdrawn until a large crystal is formed, upon completion a large solid silicon wafer weighing several hundred pounds is produced.

A salami shaped bar of silicon referred as the **Ingot** is the first step in chip making. High-speed saws slice the ingot into 'wafers' about the thickness of a Dime which is then grounded , polished and passivated to make into smooth mirror like surface. The Chip is tested first for purity.

All cell phones contains a semiconductor chip. Those chips are actually made from **Wafer**. By doping, etching or deposition wafers can be micro fabricated. Manufacturers use diamond wire saws (replacing steel wire saws) to cut wafers now. Diamond wires slice wafers much more thinly, so less material is used and wasted. Using high-tech equipment wafer properties such as i) curvature, ii) temperature, iii) thickness, iv) roughness can be maintained to standard values.

Individual solar cells get connected to make a module called PV module. A solar module with a **single solar cell** produces 0.5V (power). An industry standard module contains **36** solar cells. Cells in a module used to:

—Wired in **series** to increase current,or
—Wired in **parallel** to boast up voltage
— A **Solar panel** consists of more than one solar module
(e.g, a 330 Watt module with 72 cells) arranged in an
array called solar array.

2.2: <u>Non Silicon Thin Film Panels</u>

In this category, Solar cells are manufactured from materials

other than silicon. The products have been developed using

the term **ThinFilm Technology. <u>CdTe, CIGS, GaAs and</u>**

<u>Q-Dots (QDs)</u> all are Thin Film solar ray absorbers. Thin-

film active layers are very thin~ to be as thin as 0.1-10μm

(whereas, Si-PV is 100μm).

Thin-film (non silicon) solar cells cost less, but those aren't

as efficient as traditional photovoltaic panels. After

performing relentless research today some recognised

ThinFilms have been produced. Summarized here are major

compositions of most of the ThinFilms developed so far or,

initiated by different companies.

Summarized ThinFilms – Composition

Metals & Semiconductors	Thin Film Combination	Maximum Efficiency	Parent Company's Name
Cadmium, **Cd**			
Tellurium, **Te**	CdTe	20.4%	Firstsolar
Cupper, **Cu**			
Indium, **I**			
Gallium, **Ga**	CIGS	12.6%	SolarCity
Arsenic, **As**	InGaAs	31%	
Selenium, **Se**			
Zinc, **Zn**	CZTS & CZTSSe		SunPower & SolarCity
Sulphur, **S**			
Stannous, **T**			
Aluminum, **Al**	ZnO	<9%	
Oxygen, **O**	Al$_2$O$_3$		

FIGURE 2.1: Non Silicon THIN FILMs – combinations

- **CIGS ThinFilm & Shingles:** Thin-film made from [Ref 1] CIGS (copper-indium-gallium-selenide) uses a decade old technology and did not emerge as good as were told. [Ref.2] Japan's Solar Frontier Company is making CIGS panels yielding efficiency less than 12.6%. A new line of [Ref.3] Thin-Film Solar Shingles has been introduced in 2015. By changing panel construction from specific rare earth to semiconductor materials and with a slight modification in composition the final product is a laminated thin films–

23

hence the name "shingles." These are new to the market. Full **Criterias:** (Laminated, less efficient, less expensive, protect roof, and durable.)

● **CdTe Thin-Film**: **TOPAZ project** completed in Obispo, CA, USA in Y 2001. Huge amount of clean power produced from over 480 thousand acres of land in one of the sunniest place in Californea. First Solar is the provider company for TOPAZ solar firm that designed to create clean renewable electricity. In the vast open land sun shines there abundantly and existing transmission lines carry the generated clean sustainable energy to tens of thousands of California homes. First solar uses primarily CdTe thinfilm (efficiency >20.4%) solar cells. TOPAZ project's video link presented here.

TOPAZ PROJECT

- **GaAs wafer:** Gallium arsenide (GaAs) is a compound of Gallium (Ga, 31) and Arsenic (As, 33) is a compound semiconductor generated by reaction of group III-V elements (two semiconducting materials to absorb light at different wavelengths) with a gold backing to reflect photons back into the cell — a direct band gap generated in the composition produces the semiconductor's quality. The compound is used extensively for **microwave and IR frequencies** for **integrated circuits (IC) devices**. Methods for Thinfilm GaAs production is described [Ref.4] here. Other elements [Note-2] from gr. III-V are for films of **InGaAs or AlGaAs** etc or many other new compositions are in program. **Sharp Electronics** made extremely thin layers of a semiconductor made from Gallium-Arsenic (Ga-As) substrates. It is less expensive, too thin to work with and it works with only a single wavelength of light. Efficiency of this solar cell could reach up to 60%. Major problem is the waste heat generated from high energy liberated electron.

Bell Lab carried on research to grow thin film under the name **VBE** (vacuum chemical epitaxy). First Ga-As film was made by using TMGa (later changed to TEGa) with Arsine (AsH_3) gas. MOCVD is the most advanced deposition process. AsH_3 is used with MOCVD to clean C in a better way. Here is a link of detail progress of [Ref.5] GaAS wafer technology. **MOCVD** is the acronym for **M**etal-**O**xygen **C**hemical **V**apor **D**eposition − a technique for depositing thin layers of atoms on to a semiconductor wafer.

Knowledge of device fabrication process is a requirement to create Integrated Circuit (**IC**) that are present in everyday electronic devices. It is a multiple sequence of photolithography and chemical processing steps during which electronic circuits are gradually created on a wafer made of pure semiconducting material. For Integrated Circuit, Si along with other compound semiconductors are used with specialized applications.

TABLE 1: Si-PV vs. ThinFilm Solar Cell

Si-PV	ThinFilm Solar Cell
Purified melted SiO_2 is used	Thin Film Technology is used to generate solar cells
Silicon layer that collects sunlight is relatively thick (>100μm)	Active layer thickness of Thin films that absorb solar energy is ~ 0.1 - 10μm
Large open space is needed	Thin Film solar cells can fit practically anywhere from window to any curved space
Panels are long lasting (20 – 25Y)	Products do not last long
Cells are heavy, production cost is high too, but efficiency is high ~ 30%.	Cells are flexible, lightweight, low costs, efficiency ~ 15%.
Installation cost is high, large size panel handling is difficult.	Installation cost is low as cells are easy to handle

●**Quantum Dots** (QDs) are tiny pieces of semiconductor crystals, < 10 nm in size. Dots have different properties and characteristics. QDs are highly fluorescent nanoparticles used in medical imaging and in plasma television screens.

These are new class of nanomaterial allow solar cells to be deposited as a solution, like printing ink onto paper or painting a wall. Efficiency of QD is only 9% — however, spray on solar cells are faster and cheaper than building them from silicon.

When QD [Note – 3] is made from toxic chemicals such as Cadmium Selenide (CdSe) and Indium Arsenide (InAs) the products cost goes over million dollars for a one-kilogram bottle. An alternative approach to develop QDs was introduced by, **Ruquan, Ye** who devised a new way to make quantum dots by using graphene derived from fine black powder of anthracite coal. This product cost only hundred dollars a ton.

Researchers use QDs to make solar cells because the energy level of the semiconductor can be tuned by changing the size of the dots. Quantum dots are also used to make **solar concentrators**. At Los Alamos National Labs in USA, quantum dots are used to develop **solar window technology**.

[For more on Solar Window Technology see page# 44]

• <u>**PEROVSKITES**</u> — a unique type of <u>**THIN FILM**</u>:

A new kind of filmy material is Perovskites - available in nature in crystalline form. These are minerals made up of mostly calcium titanate ($CaTiO_3$) possesses superconductive substance. Solar cells made from perovskites are way cheaper. In 2012 and 2013 scientists were able to boost the efficiency of perovskite solar cells from just a 4% to more than 16%. These are **transparent film**, so could be easily incorporated into windows and tall buildings. **Hybrid Perovskite** is a graphene-based photovoltaic material that also emits light along with absorption. It is a high quality material and very durable under light exposure, be able to capture light particles and convert them to electricity or vice versa. For large scale solar cells, perovskite material can be modified easily to fabricate light emitting devices.

Advantage of Perovskite - Coating (micro-spray technique in use): Perovskite sheets can be made by coating using a" spray on method" introduced in University of Sheffield, England by Professor David [Note 4(i & ii)] and his research team. The efficiency of the coated product measured at Shefield was found at 11%. The spray-on technology allowed to cut down considerably the cost of Perovskite materials making solar power way cheaper even for big projects.

The spray-painting method was in use by researcher for a long time to produce thin film coatings for organic photovoltaics (OPV). The idea of replacing long chain conjugated organic compounds with perovskite material as the light-capturing layer will eventually improve efficiency of the OPV cells. Easy manufacturing process with easier to scale up production along with spray-painting method that wastes little material - these features boosted perovskite's use to obtain renewable energy.

<u>VTT</u> [Ref.6] Technical Center of Finland has manufactured roll-to-roll decorative organic solar panels (OPV) with inorganic Perovskites. Materials used there was recyclable. With rapid mass production and recyclable property, the cost of the perovskite panels dropped to a new low.

Oxford PV - Thin film perovskite solar cells can be printed directly onto silicon solar cells, or on CIGS solar cells or on glass known as Oxford PV. A 35 story office building in NYC is now wrapped with all semitransparent perovskite with **20%** efficiency-the system is going to provide 60% of its power from the windows.

2.3: <u>Organic Photovoltaics (OPVs)</u>

The sun produces 7,000 times more energy per day than what we need, but we do not know how to harness it well. A trade-off occurs between the oxidative stability and the work function of the metal cathode used for emitted electrons with conventional solar cell technology. Work function relates to the level of difficulty electrons face as they transfer from the solar cell's photoactive layer to the electrode delivering power to a device.

Organic solar cells (OPV) is built on following standard solar cell terminologies: i) light absorption, ii) electron release upon excitation which causes them to flow in certain direction, iii) Circuit close. For OPV, photovoltaic diodes primarily are combinations of polymeric organic semiconductors. An efficiency of 10% is a standard necessary for organic polymeric solar cells to sustain any commercial application. MIT (Massasatues Institute of Technology) researchers studied the behavior of electrons in many thin film organic materials and got persuaded to replace silicon panels with more affordable OPV shits. Fullerene derivatives, P3HT, Perelene diimide, Pentacene and Graphene are some of the dominant conjugated organic polymers used directly; or modified in presence of other specific linkers to use as semiconductor panel. By manipulating fabrication ideas, material scientists developed nano sized objects, these are particle sized objects between 2nm-100nm in sizes as in "nano-ink", "nanofiber", "nanorods," "nanoneedle" etc. all of these products demonstrated distinct **optical, electrical and mechanical properties.** Nanotechnology proved key for defeating cost and efficiency barriers. Nano objects are used to make solar cells to get absorb more sunlight from a wider range of angles than traditional ones. It also avoids the need for coating saving cost. OPV technology run on "spray-on coating" method as well to generate a conducting smooth surface. The polymers as nano objects be sprayed on with a micro-mechanical sprayer on plastic or ITO (Indium-Titanium Oxide) transparent conducting surface, the layer is then masked adding cathode and anode layers respectively and then ready to use as OPV. OPVs are still inefficient and intrinsic interfacial solar cells.

Schematics of an organic solar cell is presented below:

Fabricated Thin ITO Sheet

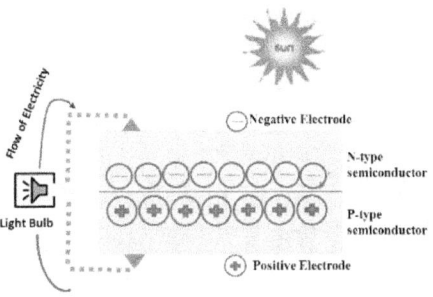

Figure 2.3: OPV Cell in Action

Details: When photons strike the n-type layer, they create **electron-whole** pairs, and the built-in electric field within the cell separates these carriers. This separation facilitates the flow of electron to the n-type layer. Appropriate wiring confirms flow of electricity for "organic photovoltaic" or OPV solar cell.

Placing an **n-type semiconductor at the top of a solar cell** is part of the design to optimize electron flow and enhance the cell's efficiency (absorption of more light generate carriers). This design helps improve the overall efficiency of the cell.

DSSC is a dominant type of OPV cell. Its production detail, polymer type used, electron transfer routs on flat surface is presented below.

●**Dye-Sensitized Solar Cells (DSSCs):** Dye-sensitized solar cells (DSSCs) rely on **dyes** that absorb light to mobilize electrons and demonstrated promising source of clean energy. Jishan Wu at the A*STAR Institute of Materials Research and Engineering and his colleagues in Singapore have developed **zinc porphyrin** and named the cell as **Dye-Sensitized Solar Cell (DSSC)**. Dyes used here to harvest light in both visible and in near-infrared parts of the spectrum. A*STAR Institute of Materials Research and Engineering Department presented a research paper detailing DSSC cell [Note. 5 (i & ii)].

DSSCs are easier and cheaper to manufacture but demonstrated **lower efficiency**. Porphyrin is a ring shaped molecule replaces Ruthenium-based dyes that have been used in traditional DSSCs for long time, with recent effort researchers developed a more efficient dye based on zinc atom with porphyrin. Solar cells using this new dye, called **YD2-o-C8** that convert visible light into electricity with an efficiency of up to **12.3%** and can also absorb infrared light.

●**Bio-Solar Photovoltaics-(a miniaturize device)**

Professor Choi, an electrical and computer engineer of Binghamton University in England developed a mini-solar cell-device [Ref.7](Lab-on-chips) for bio sample analysis. Devices that requires milli watts amount of power density to operate such as **hand-held**

Blood Analyzer or Air-Testing machine will be perfect to run with this device. The device produces good results with some types of cyanobacteria, or a combination of cyano- with heterotropic bacteria.

2.4: <u>Innovative Other Solar Technologies</u>

●**Concentrated Solar Power (CSP)**: One of the oldest technologies using solar energy on a utility scale basis is concentrated solar power or, **CSP** in short. For more than two decades **354 megawatt** Solar Energy Generating Systems at **Ivanpah Solar Power** Facility in Southern California of USA was the largest and expensive solar-power system in the world. A second 24/7 CSP solar energy service provider is the **Crescent Dunes** also in CA, USA. This one is **CSP (thermal)**, offers the potential to combine solar power with mirrors using a liquid energy storage medium. The 3rd project is in [Ref.8] <u>Dubai, UAE</u> , here a deferent technology has been applied.

CSPs - World Wide

3 CSP solar farms are in active production phase for a 24/7 service

- Crescent Dunes Power Plant, 110 MW (CA, USA)
  - Ivanpah Power Plant, 354 MW (CA, USA)
- Mohammed Bin Rashid Al-Maktum Solar Park,1000 MW (Dubai)

#Ivanpah Power Plant: Built on in CA, USA in 2007. World's largest solar plant is a glittering sea of mirrors, concentrating sunlight into three glowing towers. The plant is in the Mojave Desert opened for business in 2014. Ivanpah has 173,500 garage door-sized sets of mirrors spread over 3,500 acres of area. Each mirror has a motor controlled by a computer, which angles the reflective surface to track the location of the sun. Mirrors reflect solar rays onto three boiler towers approaching up to 40 stories high.

Work Theory: At Ivanpah, 173,500 sets of moving mirrors to the towers concentrate sunlight to generate superheat steam that used to produce electricity—each mirror is with its complicated plumbing system. The mirrors concentrate sunlight by a factor of 1,000 and direct the beam onto a tower where molten salts are heated upto 1,050 degrees F. The molten salt is stored in tanks and used to generate steam to spin turbines to produce electricity. The system supplies electricity for 15 more hours after the sun sets in with a capacity factor of 75%. Many other countries in the world have built and operating similar to CSP like plants for electricity production.

#**Dubai CSP Power plant:** A Solar powered [Ref 9] desalination plant-with a goal to produce 1000 MW of electricity by using renewable photovoltaic technology by 2030. Dubai government has adopted a plan- to construct the project phase by phase. A method to produce fresh water without burning fossil fuels for the residences in desert country has been adopted. This desalination facility is going to run by an array of solar panels and battery and resumed operation from July 2016. Plant capacity ~**13,200 gallons of**

drinking water per day for use on site. So, no burning of

fossil fuels for fresh water supply.

<u>Dubai Project: Highlights</u>

- The plant is on a plot of 285,000m^2 area
- The site is only for holding PV panels
- Project will generate 24 million kWh of electricity per year
- Save 15000 metric tons of CO_2 per year.

In many desert countries and in drought-ridden areas

desalination plants are run by using **CSP-Thermal**

technology. For long duration and cheap electricity supply

CSP-Thermal technology is one of the best. [Note 6].

How it works:

CSPs are worldwide but varies location to location. The core

idea behind "CSP-Thermal for Desalination" technology is:

i) first, to generate heat and retain that high heat by

channeling to a thermal battery (to use it in later part of night

or day), ii) second, when electricity is needed, the stored heat

gets dispatched through a **heat exchanger** to create super-

heated steam to power a traditional **steam turbine**. This part of the process is similar to any conventional fossil fuel or nuclear power plant only with zero carbon emissions and with no fuel costs to add on.

TABLE 2: PV vs. CSP System

Photovoltaic Solar	Concentrated Solar Power
Materials used here are getting cheaper day by day, PV electricity cost 6c per kWh	Price of CSP solar remain static and little pricy.
Scaling in bothways (up or down) is easy.	CSP Technology has not been updated timewise
Electricity is used where it is made.	CSP plants need huge space of open lands for production and maintenance.
Technology is easy to operate	Complex operation and difficult to maintain
Panels produce energy when the sun is up.	Advantage of CSP projects is CSP smooth out energy supply for 24/7 hrs of a day.

#CSP in **Morocco**: Morocco gets plenty of sunshine—about 3,000 hours per year, according to the Solar GCC Alliance. Taking advantage of the ample Saharan sunlight, World Bank is funding to build <u>Noor Power Plant</u> project in 3 phases in the Sahara Desert located near the town of Quarzazate. This renewable energy investment will make Morocco less reliant on imported fossil fuels, reduce million tons of carbon emission.

• **SP on <u>Road:</u>** In France, cobblestone paths and traffic-jammed streets skipped traditional bricks and pavements are now fitted with shiny solar panels on its new roadway project.

French officials announced plans to construct a **1,000-kilometer**-long (621 miles) solar roadway, with each kilometer capable of providing enough clean energy to power 5,000 homes. The photovoltaic cells can simply glued on top of existing streets and are durable enough to withstand heavy traffic and weather conditions. France won't be the first country to roll out a solar road. A 70-meter solar bike path was installed in The Netherlands in 2014 and has proved successful, [Ref. 10] with the path creating enough energy to power a house for an entire year within six months of installation.

● **SP on** <u>Trains</u>**:** Storing electricity on train is a low-tech simplicity appeal. TRAINs and rail technology to be use for it energy storage and consumption is just a clever way of repurposing existing technologies. At travel time Ttrains go up and goes down; in carrying out the task energy is involved. For smooth intermittent power production and to hold the power for when it is most needed, the train runs in **open land and a slope** at different time of day. Overall, low cost solar power is for to drive a train uphill and then letting the train roll downhill to regenerate power to reload grid when natural energy production is low on grid.

● **SP on** <u>Aeroplane</u> **(Solarstratos):** A plane fitted with 237 square feet of solar panels in its upper surface –turned to a

Solarstratos machine. The Solar-Powered Plane named

Solarstratos can witness the curvature of the Earth

- Solarstratos' is a plane that can reach an altitude of 80,000feet.
- Becoming the first plane to reach the stratosphere using solar power
- The plane is out-fitted with **237** square feet of solar panels.
- The company claim this can keep the plane in Flight over a full day.
- Solarstratos' is for commercial use and can seat two person.
- Company revealed the plane to public on **December 7, 2016**

- **Solar Power in <u>SPACE</u> (SBSP):** Inhabitat -

Cleantechnica - SMA are pioneer in exploring space for

energy source. SBSP is the concept of collecting solar power

in space using a Solar Power Satellite (SPS). **NASA**

launched **1 GW** wireless microwave orbital solar system

reflectors for a **space-based solar power** (SBSP) and

positioned in orbit that unfurl and direct solar radiation onto

solar panels. The energy is converted either into a laser or

microwave beam and transmitted to receive stations on earth

that convert it to electricity and feed into electric grids. Microwaves work in all weather, giving them an advantage over lasers that are absorbed and scattered by water molecules in clouds. A very expensive endeavor.

- **SP on <u>Water</u> (Floating Solar Panel)**: Countries like India, and Japan lack open free space to install huge arrays of solar panels. A floating solar plant built on numerous canals and lagoons of water bodies is used to harness solar power. India planned for 10MW solar power, Japan sets goal for 70-MW plant with floating solar panels. By building the plants countries saved both cash and valuable real state. In addition, no change in the ecology of the water while reducing evaporation from the water pots persevering water levels during extreme summer. Panels installed on floating platform get cool faster.

- **SP on <u>Windows</u> – Solar WindowTechnologies (SWT)**: This technology has introduces a transparent coating on glass or plastics for electricity generation. The vision is for able to generate renewable electricity through windows from all four sides of skyscrapers in a city. The electricity generating technologies is made from **organic earth-abundant materials**, making it different from the plant based photosynthesis process that produces chemical energy. The Solar Windows to be connected using the same panel based wiring system but the connected wires will remained invisible to general eyes. The other benefit of SWT is: coating can power a 50 storey building with no distortion anywhere, whereas with conventional panels it would require six to eight acres of land to hold the solar panel arrays to power the same building.

CEO of SWT explained; sun light when hits the coating on windows, the mobility of the electrons activated, producing electricity.

Video link of SWT explained by a professor

SWT technology differ from roof top or other solar sources because the solar windows can be integrated into large or small buildings. Clean electricity is generated on transparent glass tinted in a high-demand color and framed in aluminum, popular with architects and developers of commercial towers. For high speed, high volume, roll to roll, and sheet-to sheet SWT product is today's green energy solution . The original SWT video link is here transparent power of a working Solar Window

• **SP as <u>Sunflower</u>:** Solar Sunflower, is a Swiss invention developed by **Air light Energy, Dsolar** a subsidiary of

Airlight, and IBM Research in Zurich, introduces a new product called **HCPVT** to generate **electricity and hot water** from solar power. HCPVT stands for "highly efficient concentrated photovoltaic/thermal," and has reflectors that concentrate the sun rays—many times. These highly efficient photovoltaic cells are capable of converting concentrated solar power into electricity, without melting in the process. Airlight/Dsolar provided Sunflower's reflectors and superstructure. IBM supplied the photovoltaics. The specialty about the Sunflower is that it combines two technologies—CSP/ thermal with PV solar power— to attain higher total efficiency. Each Sunflower has six "petals," and **each petal holds six reflectors made from aluminum foil**. At the focal point of the **36 reflectors** there are **six collectors**, one for each block of reflectors. Collectors are an array of **GaAS** photovoltaic cell efficiently convert sunlight into electricity (38% conversion rate). One Sunflower can generate **12kW** of electricity in total in a day. The GaAS cells are pricy and the CSP concentrated to a single point causing a very high temperature. [Ref.11] IBM solved the problem with its hot-water-cooling technology. (click the link above for detail information).

●**CSEM Photovoltaics** ('Invisible' solar panels from Swiss firm): A Swiss CSEM (Swiss Center for Electronics and Micro technology) Firm, has developed recently the world's first truly "invisible" solar panels making solar an aesthetic commodity.

Work Theory: CSEM solved the problem of absorption by focusing on solar technology that absorbs light from outside of the visible spectrum. These solar modules are very sensitive to Infra-red light. SCEM [Note 7] combined the silicon based panels with a **special filter**. The filter was synthesized by using innovative nanotechnology. When sunlight strikes the module, the filter scatters the whole visible spectrum while transmitting infrared light. As the light beat down, they knock the electrons off the silicon. The negatively-charged free electrons preferentially created an electric voltage on one side of the panel that can be collected and channeled. These panels can work at temperatures of 20 to 30 degrees C below standard models, because visible reflected light will not produce any heat. The panels could also be install on cars, buses or planes without interrupting the look of the vehicle surface.

YouTube video demonstration of CSEM has been included here.

Solar In **INDIA**: India, with an aim of increasing its solar power capacity to [Ref.12] 100GW by 2022. It also joined to an International Solar Alliance with 120 countries at the

COP21 Climate Conference in **Paris**, on December, 2015.

- o World's first solar-powered airport was opened in the city of Kochi in August, 2015;

- o A plan for solar powered train to roll out by the Indian Railways,

- o India would be the country for world's largest **750 MW solar power station** in the central Indian state of Madhya Pradesh by 2017.
- o Punjab holds the world's largest **rooftop solar** power plant.

Punjab project - Highlights:

- Eight rooftops across 82 acres,
- Produce **11.5MW** of electricity,
- Costs Rs 1.35 billion,
- Reducing 400,000 tonnes of carbon dioxide (for 25 years).
- Enough electricity to power 8,000 Indian homes.

Kamuthi Solar Farm [Ref. 13] in India has opened new solar project in Kamuthi, Tamil Nadu in India. It is now one of the biggest in this region. A 648-MW solar energy facility beats the previous title holder, Topaz Solar Farm located in CA, USA.

Kamuthi project - Highlights:

- The site stretched 2,500 acres (10 square Km) of area.
- Equivalent to 60 TajMahals
- With 2.5 million solar modules used to convert to electricity
- The Company that steered the technology here is Adani Power built it in 8 months.
- Save enough greenhouse emitting fuel-means saved poor people in southern India

Solar Energy in **CHILE**:

With major investments in renewable energy, Chile is considered as an incredible solar energy source. The country produces so much SE that it's giving it away for free. Solar capacity from the country's central grid has increased fourfold to 770 megawatts since 2013. Another 1.4 gigawatts will be added in coming years with many solar power projects under development. Chile now has 29 solar projects, another 15 are on the way. Cost of solar is absolutely nothing for certain regions in recent months. Largest solar PV project commissioned to **Enel Green Power Chile Ltda**. The best solar resource in the country resides in the **Atacama Desert** [Ref. 14] in the north. Acciona SA, Spain's second-largest clean-energy company, will invest $343 million to build a 247-megawatt solar farm in Chile's Atacama Desert. (watch the video link of [Ref.15] Acciona project presented in Ref. 15.)

TABLE 3: All Panel Types – Surmised Info.

Panel Comm. Name	Compo.	Unit price, $	Effic. (%)	Developer
Monocrystalli ne	Silicon in pure form	<1.4 per W-h	15-18	Sun Power
Monocrystalli ne back contact	Silicon with Cu as back contact	1.4	21.4	Sun Power
Multi-crystalline	Si Ingot	0.50/Watt	14-16	Yingil Green Energy, China
Perovskite	CaTiO$_2$	cheap	16	Oxford, UK
Hybrid Perovskite	CaTiO$_2$ + Graphene	cheap	11	Sheafield, UK
ThinFilm CdTe	Cadmium Telluride	1.0	14	First Solar
ThinFilm CIGS	Cupper, Indium, Germanium, Selenide	cheap	12.6	Solyndra, USA
Ga-As (Ultrathin)	Galium Arsenide	cheap	60	Sharp Electronics, Japan
2 Junction solar cell (Au backing)	[(Ga, I, P) (Ga, As)+Au]	cheap	31.1	Sandia, NREL
Dye Sensitised solar cell; YD2-o-C)	Zinc poroyrin with Ru	way cheaper	12.3	Singapur, NREL
Carbon based polymer	Fullarene + polymer	cheaper	10.8	North Carolina, NC
Carbon nanotube with Graphene	Graphene+C -nanotube	cheaper	-	Switzerland
Ultra-Thin OPV as CPZ	C60	-	-	Amherst, USA
Hybrid (organic+inor ganic)	PbSe + Fullarene	-	15	Chicago, IL
CSP & CSP Thermal	Mirror & Salt	cheapest	~	Sandiago, CA

CH – 3

SOLAR ENERGY TERMS

Solar power holds the theoretical potential to solve world's all the energy problems. However, harnessing this nearly limitless, clean, and renewable power for our choice of use is still difficult. To produce solar energy anywhere: for sure sun to shine the panels, through the glass, films, or coating and the underlying solar cell converts the light into electricity. Devices used for this purpose are always: panels, metallic racks, master cables and clips, inverter (DC → AC current converter) meter box etc. Battery for storage is essential for "off the grid" service and opt out for "grid" connections. When customers send back excess electric charge back to the grid in exchange for incentive for rooftop residential system, battery turns optional to the system. This policy is termed as Net-metering (NEM). In this chapter we are providing definition of some essential terminologies and describe topics and terms of solar energy essentials.

Energy Basics

Two equations are used extensively for solar energy conversion process:

[Electric **Power** unit is Watt for every relationship]

$P = I^2R$, Power = (Ampere)^2 x Resistance [Note 8]

Power is equal to the single variable of current "I" squared times the Resistance (R) along the path of the circuit. Power consumption can be minimized in the circuit keeping R small.

Resistance in electrical wires consumes power. Power consumption increases, when current passing through the circuit increases. Maintaining higher voltage at two ends of a circuit current consumption can be reduced. Low power consumption eventually make electrical system efficient.

Low current justifies low power consumption saves energy.

Applying Ohm's law to substitute the I term with the

derived relationship ended up to:

$P = V^2/R$, indicates that under identical condition it is

possible to get same **power** with a low current input by

changing voltage at point A to point B. Overall, double the

current to power double; or double the volage which doubles

the power again.

Power to Energy: Conversion

POWER

Unit ---- Watt (W)
The smallest unit of Power is --- kW
James Watt introduces (horsepower [hp] unit)
in 1780
1 hp = 746W ~750W
1 kW = 1.33hp

ENERGY

Work done- or
Energy expended (by a force) in a shorter time
Unit of Energy is kWh
[This energy unit is use for an EV battery pack]
To relate Power to Energy, duration (time) is the
essence

POWER & ENERGY

kW ⟶ kWh (linked through Time)

Scientific unit of Energy is Joule (J)
[A Force of 1Newton on 1kg of mass moves 1m]
If the same work done in 1sec, that is POWER
1 J = 1 w – s

kW: stands for **kilowatt**, a unit of energy used to measure electrical Power.

kWh: The acronym for **kilowatt-hour**, the amount of electrical energy consumed when 1,000 watts of power used for **an hour**. Also, 1kWh means an energy source supplies 1000w (1kW) of energy for 1 hour constantly. On average, a solar energy system provides radiation for 5h/d. Therefore, if a system of 1.8 kW unit will generate energy for 5h/d, for 365 days/y is totalled to:

$$\equiv 1.8\text{kW x } (5\text{h/d}) \text{ x } (365\text{d/y}) = 3{,}285 \textbf{ kWh/y}.$$

Because, Inverters are only 90% efficient now, therefore the expected energy output from the system is: 3,285kWh x 90% = 2956.5kwh energy to use.

Energy Output Calculation:

$$E = a*r*h*PR$$

E = Energy generated
a → area,
r → solar radiation (η*1000W/m²)
h → duration,
PR→ is for system loss|

Solar panel specifications are quoted in terms of the theoretical electrical output potential such as X-Watt. Let's consider a solar panel that produces 8-10W/m². A roof with an area of 200m² if covered with these types of panels, will produce: $(9W/m^2) \times 200m^2 = 1800W$ or, 1.8KW. The system will then termed as 1.8KW power system.

• **Grains:** Solar cell efficiency (silicon PV or ThinFilm) is somewhat related to grain boundary which is the interface between two grains or crystallites in a polycrystalline material. Grain boundaries are tiny defects that normally act as roadblocks to efficiency, inhibit carrier collection which greatly reduces solar cell power. "Larger grain sizes mean the crystals on the film is more continuous, and the electrons passing through the film face fewer interruptions" said [Ref.16] Sonia Ruiz-Raga. Thin film CdTe solar cell's recent research justifies that atom-scale grain boundaries involved in this type of cell enhance performances. For CdTe cells grain boundary structure control is the new direction that helped raise the cell efficiencies to close to theoretical maximum. Currently, the recorded **CdTe cell efficiency is 20.4%** compared to **Si PV of 22.4%.** Researchers' think that if all the grain boundaries in a thin film material be aligned in same direction, it would help improve cell efficiency even further.

●**Polysilicon panels**: The initial step in producing polysilicon panel requires melting silica rock at 3,000 degrees Fahrenheit generally using electricity from coal fired plants. This is the primary reason for why panel price turns so high. Once, polysilicon (several fragments of silicon are melted down it then recrystallized to make wafers of polycrystalline solar panels that upon exposure to sunlight generate electrons. World's second-largest supplier of polysilicon is US. US sends polysilicon to China, which returns to US as panels and modules — as finished products.

●**Mono-crystalline** silicon panels – When silicon is in more pure form is used to make mono-crystalline solar panels yielding higher efficiency and low cost per W-h of energy.

Adding **nano needles** to (**black**) monocrystalline silicon enables solar cells to absorb more sunlight from a wider range of angles than traditional ones. At the same time, it avoids the need for anti-reflection coating that saves costs of the panels.

Mono crystalline silicon is usually doped with (P, N, or As) to yield **n-type silicon** to use as semiconducting substance to produce more efficient solar cell. In contrast, p-type semiconductors doped with Boron (B) for enrichment. Research suggests p-type silicon shows higher degradation rates over time than the n-types. SunPower Company chooses n-type materials for solar panel manufacture to provide higher efficiency. Sun Power's construction differs greatly from its competitors and it is complex.

●**Diode, and Semiconductor Diode with p-n junction:** A diode is an electrical device allows current to move through it in one direction only. It has low resistance (~0) to current

in one direction and high resistance (~∞) on the other.

Figure 3C: Diode Graphics.

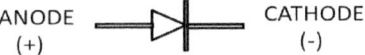

Semiconductor diode is the most common type where two

crystalline pieces of semiconductor materials with a p-type-

connected with an n-type in electrical terminal. Most diodes

today are made of Silicon, but other materials used are

Selenium, Germanium and Indium. An N-type silicon is

doped chemically with phosphorus (P) (having 5 electrons in

its outer most shell), one electron become "free" and moves

easily within the crystal lattice when voltage is applied. On

the contrary, a p-type silicon is doped with Boron gas with 3

electrons in its outer shell of the atom. This leaves one

silicon atom with a vacant spot in its outer shell (a "hole")

that readily accepts one electron. P-type silicon is positively

charged.

•Silver paste for solar panel: 90% of crystalline silicon solar cell use silver paste. Silver has the highest electrical and thermal conductivity of all metals, and it's the most reflective. When sunlight hits the solar cell it generates electrons. Silver paste facilitates conduction by collecting these electrons orderly for electric current and then transports electricity out of the cell for distribution. The conductive nature of silver enhances the reflection of sunlight to improve the energy to be collected. An average solar panel actually needs about two-thirds of an ounce (20 grams) of silver. At around $20 an ounce it contributes ~60$ per panel as solar cost. Silver is now one of the primary ingredient for photovoltaic cells.

• ROBOTs (for panel installation) – Installing Solar Panel (roof-top) is getting easier day by day as barriers accessing SE is wiping out constantly. Physical labor is needed to install panels on site – ROBOTS are getting trained to take over this task.

Trained Robots are to replace human solar installer. Once, perfection come in Robots' skill, the Robots will be placed appropriately to install panel- as a result ultimate installation cost would cut down many times when this become a reality.

• **TRIFECTA-** TESLA Motor Inc.'s name is associated exclusively for humanizing solar electricity storage technology. The company's mission is **harness, store, and power**. Home battery line like Powerwall, Powerpack– is available in the market now. Powerwall is for to store and distribute electricity in consumer's home. TESLA introduced **TRIFECTA** idea as a solution to accelerate transition to sustainable transportation by adopting Powerwall and Powerpack in the system. Trifecta idea demonstrate a joint service from Solar panel on roof top, TESLA electric car in a driveway, and Powerwall in the garage.

Figure 3d: TRIFECTA

• **Microgrid** is an electric grid that is much smaller than a city, state, or national grid and contains both generating assets as well as energy demand sources. Microgrid can create a self-contained energy ecosystem.

Microgrid is good for distributed solar on rooftops and also ground solar generation. The main electric grid is likely to feed into a central point that would control all communication points of the system including feeding energy to the microgrid itself. In some locations, microgrid operate independently of a central electric grid and smooth out demand from the larger grid. The first two companies to build microgrids are SolarCity and SunPower. **Siemen**s is a large micro-grid supplier company.

• **Grid Sharing** is when a given grid system becomes saturated with solar, it can connect with nearby grids to ease the pressure off from the lines. For example, California (CA, USA) and six (6) neighboring states in the west coast of

USA (NV, OR, UT, AZ, CO) has a grid sharing agreement and running an interconnected profitable energy business inside USA.

●**Grid-parity**: Grid parity, is a term used for when an alternative energy source can generate electricity at a cost that is less than or equal to the price of purchasing power from the grid. Put simply, it will cost no more in the long run to use solar panels for energy production than to buy electricity from utility companies.

●**Net Metering**: An alternative to power storage for renewable energy is **NEM** that allows residential solar customers to sell their excess solar electricity back to utilities. Free renewable energy is evolving continuously, phase by phase. Rules and policies are also playing out stepwise.

Solar panels installer (during buying) get full credit upon agreement that excess energy will be feedback to the electric grid. The benefits goes to solar home owners who gets full retail credit for that electricity is at (~8C /kWh).

- **kWh Batteries** are for electric energy storage are on demand now to integrate renewable energy for power generation for higher capacity and for long lasting objectives. A similar innovative solution developed by **Lancashire, UK Company** for storing and distribute solar energy during the day using **titanate lithium-ion** battery. This high performance Lithium-ion titanate batteries have a very long life, with approximately 15,000 charge-discharge cycles, compared to the usual 3,000 cycles for traditional batteries. A ceramic separator is used here to increase the safety for the key elements. The storage unit can store up to 500 kWh of energy equivalent to the production of electricity from 2,500 m^2 area of solar energy. **TESLA's PowerWall 1 (Powerpack-1** for businesses and utility initiatives) is powerful but expensive endeavor for consumers.

Powerwall is TESLA Motors Inc.'s home electric storage, can store up to 6.4 kWh units of charge of electricity from home solar systems and provides backup during outage. It is a 4 feet tall metal box, weighing 214 pounds with retail price 3000$. With an accessory of a bi-directional inverter converts direct-current electricity to use (in our dishwashers and refrigerators needs). Inverter is pricy therefore it adds up money to the retail price of Powerwall. The higher price discourage customers to own the power storage and slowed market expansion for TESLA. A link is provided about detail description on Pewerwall & Powerpack (published earlier by this author in author's webpage) [Ref.17] here.

TESLA's Powerwall is a convenient energy storage till date mounted on wall is basically a lithium ion rechargeable battery. At least **seven (7) Si-PV solar panels** are needed on the rooftop to recharge powerwall. Tesla powerwall is now expanding its business in countries such as Australia, South Africa and Europe.

●**Solar Orb** is a large high power device and in direct

contact with the sun at all times tracking sun's position in the

sky using sensors and equipment. The device is named as

Rawlemon contains an Ref. 18 eyeball-shaped lens that uses refraction to concentrate sunlight. German architect Andre Broessel designed and uses a large glass sphere lens, whch collects diffuse light from multiple angles from sun. The device works like a magnifying glass and around 70% more sunlight is possible to collect than traditional PV panel. The installation costs is ~$6000 for Rawlemon's XL brand but the price is getting down yearly.

- **Solar by Artificial Photosynthesis (AP):** Photosynthesis (a natural sunlight conversion process), relies on pigments like chlorophyll on leaves to capture the sun's energy and the process involved CO_2 and H_2O turning the compounds into energy as sugars. The natural process carries on each and every green part of a plant in nature. John Markoff of

Sanfrancisco, CA, USA used appropriate wordings to describe it better:

*"Subatomic properties of atoms are used by plants to turn sunlight into photosynthetic energy in a few **million-to billionths of a second**."* This recognition should be the key for us to extract solar energy.

To simulate the way plants use sunlight to turn carbon dioxide in the air into energy (glucose) and oxygen — is ideal to create solar cells by using either **OPV** instead of **PV** or **hybrid OPV-PV films.** Use of **Perovskite** films for cost effectiveness of the process is under consideration. For sustainable energy for the world 'photosynthetically solar energy generation' idea is one of the best choice is my opinion.

CH - 4

UNDERSTANDING THE POWER OF THE SUN
For Solar, By Solar, To Solar

Tesla Electric car maker, Elon Musk is Tesla's CEO joined to a podcast **episode** of "The Joe Rogan Experience (October/23)" where he said 'there's a great energy source that can fulfill all our needs right above us: **the sun**.' Musk explained further by saying, "the sun is converting more than four million tons of mass to energy every second and requires no maintenance. That thing just works. We have **a giant fusion reactor in the sky."**

Fusion in the sun is the process of combining hydrogen nuclei into helium, releasing large amounts of energy. This fusion happens in the core of the sun, where the temperature is millions of degrees. Thus, the source of the sun's heat is mostly the nuclear fusion occurring at its core level. The fairly sharp outline (contour) of the sun is actually the surface where the photons produced by the nuclear fusion reactions deep inside the sun scatter incessantly. "From that surface called photosphere the photons fly towards us, unhindered. But this photon scattering prevents us from seeing directly inside. The interior of the sun is opaque to particles of light not transparent," said a physicist at Max Plank University, Professor Chitta, L.

The Sun in our solar system is the only star that we can study up close. By sending star skimming spacecraft such as the Parker Solar probe [Ref. 19] the flyby that can reach at its closest of the Sun. At this closest approach, the spacecraft will come within about 3.9 million miles (6.2 million kilometers) of the Sun. This is the outermost part of the Sun's atmosphere (called corona) . Brutal heat and radiation prohibit any probe/spacecraft to go closer to it.

The only accessible location is within the orbit of planet

MERCURY's orbital path.

Our understanding of other STARS comes from studying our Sun:
What we do know now about stars is that:

All solar type stars have their own cycles

Our Sun takes 11 years to cycle

The star's convection motion causes the magnetic field to
cycle between periods of high and low intensity   Flow of magnetic field

A Star's cycle is also related to its rotation rate and brightness

Faster spinning, but fainter stars have longer cycles

This chapter presented simply to get to know more about

the Sun (in most simplistic way) and its various solar

constituents, such as charged particles and magnetic fields.

The term '**SOLAR**' is used to describe things relating to the

SUN. Solar power is obtained from the sun's **light and heat.**

Radiation causes chemical reactions, or generate electricity.

The total amount of solar energy received of on Earth is

vastly more than the world's current and anticipated energy

requirements.

Google defines the word 'solar' as: of, derived from, relating to, or caused by the sun measured by the earth's course in relation to the Sun (which is very much similar to voicing the phrase: **'For Solar, by Solar, to Solar')**. Such is the delight and serenity of human experiences, solar energy makes a comfortable backdrop for all from living to dead creatures alike. Like it or not Sun is real, the super active force behind every life on the planet Earth and beyond?.

Scientists are now busy discovering more about the sun (by sending solar probes), their focus is on three (3) defined iteams: the Photosphere,[Note 9] the Corona, and the Solar Wind.

> Precisely how the sun generates the solar wind is still remained unclear to physicists.

The phenomenon of superheating in the corona region puzzled solar scientists very much. Collecting data from different solar probes, scientists are trying to solve this puzzle. That the surface temperature of the Sun at its photosphere is approximately 6,000°C, whereas, within the corona, temperatures can soar to a surprising million degrees and beyond. The physicists are searching answers for: why our star's outer atmosphere is over 200 times hotter than the underlying surface?

'T' conundrum: The surface temperature of the Sun at its Photosphere level is approximately 6000 °C, whereas, within the Corona region the temperature can soar to a surprising million degrees and beyond.

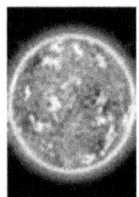

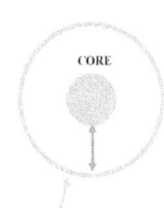

The sun is a ball of gas, so it doesn't really have a surface

The Photosphere is around 3,100 miles (5,000 km) closer to the core

The Corona extends around 5 million miles (8 million km) from the Sun's surface

Solar Wind, Photosphere and Corona are shown above in the images.
The measurements are NOT in scale.

Sunlight Distribution

● **Sun Light** (analysis): The fraction of Sunlight that reaches on the surface of the earth every day comes in the form of Electro Magnetic Radiation (EMR). Part of Sunshine that we enjoy most is called the **white light** consisting of all different wavelengths (wavelength of very high to very low) of light. Visible light is one subset of EMR. Spectroscopists marked two ends of EMR array validating two distinct color zones, from red to beyond red end is **IR/Microwave** region, and the opposite end is purple to beyond purple known as **UVB, X-ray and Gamma** ray region. Distinguished physicists' like Max Plank, Einstein, and Heisenberg and many others have studied and defined white light as "quanta," containing energy packet possessing particle or, waves based on any experimental setup. Put simply: sunlight reach on earth's surface as EMR containing energy pack known as photons, and one subset of EMR is visible light that we are able to see through (image below.)

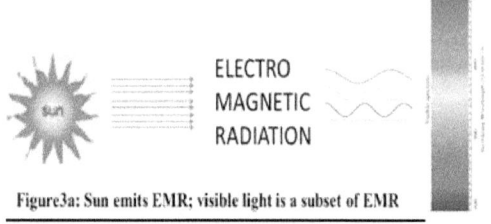

Figure3a: Sun emits EMR; visible light is a subset of EMR

Visible Light is a tiny part of ELECTROMAGETIC RADIATION

A better wording of expression this phenomena is: EMR consists of "energy packets" called photons that behave as "wave or particle" are of dual natured entity. High energy photons produces blue light (UV), low energy photons make red light (IR). Photon energy from sun light reaching on any material surface causes electronic excitation within the molecule; this conduct changes in bonding or chemistry of a molecule as a whole. Source of energy of this phenomenon termed as 'Free energy.'

Since 1950 scientists & engineers' ongoing effort establishes today's operational form of solar energy industry.

Still today we are far from achieving on average **10%** of solar energy for our livelyhood. The goal set is for **100%** carbon free energy. Some cities made progress, but some localities are still struggling to embrace this new form of solar technologies. Many companies are already helping people to harness the power of the sun. Because, 'It's the public anywhere whose **enthusiasm** make successful to any initiative or new technologies to adopt,' meaning public engagement is essential for renewables to become a mainstream energy source.

ACKNOWLEDGEMENT

I, as an author wishes to thank everyone whose help made this book possible. This book would not have been possible without that timely help. I extend my sincere thanks to my solar research supervisor, Dr. Hemali Rathnayake whose guidance motivated me most to learn solar energy technology. Eventually, Dr. Hemali managed an opportunity for me to work in an ongoing 'Organic Thin Film Develop' research project @ the WKU university's Material Characteriziation Labrotary. The project was OPV related. I am indeed grateful for this research.

The relevant Online articles that I've used and those were huge in numbers — their primary (#1) source were PV Magazine. I deeply attached with this source. Some other data that has been used to compile the subject matter of the book was all sourced from reliable websites including: U.S. Department of Energy (gov.) and U.S. Department of Energy, Environmental Protection Agency and so on. Details about many inner facts I came to know from professors and technicians who were professionals @ WKU lab. Finally, in publishing the book as Online, I've benefited enormously from the Amazons'/'Self Publish with us' - link. At the end, my sincere thanks is also for all those professionals and resources for extending the overwhelming opportunities that I've utilised @ WKU (without any fear or threat). I am so so so grateful to all of them.

NOTES

Russel Ohl studied the conductivity and properties of
various types of crystals. His work with semiconductors and
P-N junction led to items such as:
- The Transistor,
- LEDs and
- laser Diodes.

1. Bell Labs in 1940: John Burden, Walter Brattain and
   William Shockley- invented transistor, a tiny device
   that could direct and amplify electrons. The transistor
   was built around a material called a **semiconductor**-
   which could send electric currents in one direction,
   but not the other. In time, silicon would become the
   preferred semiconducting material, and the tiny
   devices that resulted would become known as
   "Chips." In replacing bulkier, less reliable vacuum
   tubes, the transistor became the basis for all
   electronic devices, allowing scientists and engineers
   to make ever –smaller gadgets or, transistor radios
   that fit in a pocket or, in one's hand. Eventually,
   companies made all new kind of equipment,
   semiconductor and gadgets with increasing abilities
   and ambitions opening door of innovations.

2. Semprius (pioneer in GaAs) designed small Ga-As
   cells. Tiny cell size produce so little heat that they
   don't require cooling, which brings down the cost of
   the solar cell. A small black square of Ga-As on each
   cell is the semiconducting material that keeps costs
   down by reducing the size of Ga-As film. John
   Rogers is the lead author of carrying this research
   successfully publishing several authenticate reports in

scientific journals. John Rogers see a great deal of potential in the future and developed cell design by simplifying steps.

3. "The QD method is simple," Ye, the researcher described; "I place this in a solution of acids for one day, then heat the solution on a hot plate." By tweaking the process, Ye can make the material emit various light frequencies, creating dots of various colors (red, blue, yellow, violet) for differentiated tagging of tumors. Ye's dots are coal-based dots compatible with the human body.

4. i) Prof. David Lidzey and colleagues at the University of Sheffield, England. (Deposited perovskite using spray heads could significantly reduce manufacturing costs. Jon Griffin/University of Sheffield) with them were the pioneers.

ii) Recent Perovskite news: A new research from the National University of Singapore shows that the compound, Spiro-OMeTAD is the key material for perovskite playing important role in transferring energy from the solar surface to the inner electrode. A new way to crystalize Spiro-OMeTAD enabled the researchers changing the efficiency of perovskite to a new high (18%). The new material's crystal structure's role suits the pathway to enhance efficiency of the cell.

5. i) Journal of the American Chemical Society 136,

265–272 (2014). DOI: 10.1021/ja409291g

Luo, J., Xu, M., Li, R., Huang, K.-W., Jiang, C. et al. N-annulated perylene as an efficient electron donor for porphyrin-based dyes: Enhanced light-harvesting ability and high-efficiency Co $^{(2+/3+)}$ based dye-sensitized solar cells.

ii) Earlier, this research team developed **DSSC** known as **WW-5** and **WW-6** combines a zinc porphyrin core with a system of fused carbon rings bridged by a nitrogen atom, known as an N-annulated perylene group. Solar cells containing these dyes absorbed more infrared light than YD2-o-C8 and had efficiencies of up to 10.5 %. Porphyrin and perylene sections of these dyes are connected by a carbon–carbon triple bond all of these act as an electron-rich linker, improved the flow of electrons between them. This bond also reduce the light energy needed to excite electrons in the molecule, boosting the dye's ability to harvest infrared light. These chemical groups are bulky, dyes solubility and aggregation—have been improved.

WW-5 and WW-6 are slightly less efficient than **YD2-o-C8** in converting visible light into electricity, and also produce a lower voltage. Modifications based on the chemical structure of the dyes simplified the problem and is currently under consideration for scale-up with the researchers.

6. Desalination plants can also operates through advanced technologies called Reverse Osmosis (RO). Plants that use RO, pushes salt water through polymer membranes that trap salt ions while allowing water molecules to pass through. This is an energy-intensive process. Plants that use heat generated by concentrated solar power arrays to distill seawater into fresh water is comparable in cost and output to some grid-powered desalination plants but not the

high cost RO technologies. However, RO backed by grid electricity is an option in oil-rich Middle East.

7. **CSEM** develops a technology for integration to buildings. Silicon PV solar modules are usually of blue/black color with a monotone hue. Now, solar modules can be integrated and customize everywhere.
https://www.youtube.com/watch?feature=player_emb edded&v=d0a_A9E40bQ

https://youtu.be/d0a_A9E40bQ

8. a) P = I x V [Watt =Amp. x Volt.] → I = P/V
b) V = I x R [Volt. = Amp. x Resistance] → I = V/R
Substitute the V term in equ. a) from equ. b) leads to:
P = I²R, Power = (Ampere)² x Resistance.

9.

Image: The Sun and the Earth

Scientists are now busy discovering more about the sun (by sending solar probes), their focus now is on three (3) defined iteams: the **Photosphere, the Corona, and the Solar Wind.**

- The PHOTOSPHERE: Considered roughly as the sun's surface. The Sun's photosphere, holds a nearly uniform temperature, since all photons radiating from the surface have been exchange heat through interactions to its interior from here.
- CORONA: The Sun's upper (outer) atmosphere is known as the corona consists of charged particles and magnetic fields. The corona's estimated temperature exceeds 2 million F (1.1 million C), while the photosphere is around 10,000 F (5,500 C).
- SOLAR WIND: Solar wind—the outward flow of charged particles from the Sun that influences all planets, including Earth—by circling closer to the solar surface. The solar wind consists of plasma - ionized gas, or gas in which the atoms lose their electrons - and is mostly ionized hydrogen. "Solar wind is ejected outward into interplanetary space. A key challenge to phusicist is to identify the dominant (what) physical process that powers the solar wind," said Chitta, L. (a physicist at the Max Planck Institute for Solar System Research in Germany.)

The relentless high-speed flow of charged particles from the sun fills interplanetary space. On Earth, it triggers geomagnetic storms that can disrupt **satellites** and it causes the dazzling auroras - the northern and southern lights - at high latitudes.

REFERENCES & LINKs

1.https://en.wikipedia.org/wiki/Copper_indium_gallium_selenide_solar_cells

2.https://en.wikipedia.org/wiki/Solar_Frontier

3.http://www.solarpowerworldonline.com/2015/01/shedding-light-solar-shingles/

4.https://phys.org/news/2010-05-method-gallium-arsenide-solar-cells.html

5. https://fbsolarllc.com/2020/06/04/gaas-thin-film-how-much-we-have-progressed/

6.https://en.wikipedia.org/wiki/VTT_Technical_Research_Centre_of_Finland

7. http://www.elveflow.com/microfluidic-tutorials/microfluidic-reviews-and-tutorials/introduction-to-lab-on-a-chip-2015-review-history-and-future/

8. http://ecowatch.com/2016/05/04/worlds-cheapest-solar/

9. http://www.apricum-group.com/dubai-shatters-records-cost-solar-earths-largest-solar-power-plant/

10. http://www.aljazeera.com/news/2015/05/150510092535171.html

11. https://newatlas.com/ibm-sunflower-hcpvt-pv-thermal-solar-concentrator/33989/

12. https://www.pv-magazine.com/2016/11/09/adani-completes-100-mw-punjab-solar-farm-in-india_100026849/

13. http://www.aljazeera.com/news/2016/11/india-unveils-world-largest-solar-power-plant-161129101022044.html

14. https://cleantechnica.com/2016/12/06/chile-announces-plans-1-gigawatt-solar-park/

15. https://twitter.com/ACCIONA_EN/status/687231158418026496

16. https://www.researchgate.net/profile/Sonia_Ruiz_Raga

17. https://fbsolarllc.com/2020/06/04/wish-to-have-a-backpack-full-of-power-source/

18. http://www.rawlemon.com; www.igg.me/at/rawlemon-transparent-power-generators

19. **Mrigakshi Dixit**. "NASA's Parker probe rings back home after historic sun flyby at 430,000 mph" provided by Interesting Engineering (IE). January 3rd, 2025.

https://www.msn.com/en-us/news/technology/nasa-s-parker-probe-rings-back-home-after-historic-sun-flyby-at-430-000-mph/ar-AA1wTDIt?ocid=socialshare

A BONUS Link:

[This Youtube Video is on Solar Panel Manufacturing technologies]

https://www.youtube.com/watch?v=alQFVKYLwT0
How Solar panels are made

THE END

www.ingramcontent.com/pod-product-compliance
Lightning Source LLC
Chambersburg PA
CBHW070104210526
45170CB00012B/740